SOURCE—THE MULTILATERAL PLATFORM FOR SUSTAINABLE INFRASTRUCTURE

DECEMBER 2023

ASIAN DEVELOPMENT BANK

ADB

© 2023 Asian Development Bank
6 ADB Avenue, Mandaluyong City, 1550 Metro Manila, Philippines
Tel +63 2 8632 4444; Fax +63 2 8636 2444
www.adb.org

Some rights reserved. Published in 2023.

ISBN 978-92-9270-519-0 (print); 978-92-9270-520-6 (electronic); 978-92-9270-521-3 (ebook)
Publication Stock No. TIM230581-2
DOI: http://dx.doi.org/10.22617/TIM230581-2

The views expressed in this publication are those of the authors and do not necessarily reflect the views and policies of the Asian Development Bank (ADB) or its Board of Governors or the governments they represent.

ADB does not guarantee the accuracy of the data included in this publication and accepts no responsibility for any consequence of their use. The mention of specific companies or products of manufacturers does not imply that they are endorsed or recommended by ADB in preference to others of a similar nature that are not mentioned.

By making any designation of or reference to a particular territory or geographic area, or by using the term "country" in this document, ADB does not intend to make any judgments as to the legal or other status of any territory or area.

Please contact pubsmarketing@adb.org if you have questions or comments with respect to content, or if you wish to obtain copyright permission for your intended use that does not fall within these terms, or for permission to use the ADB logo.

Corrigenda to ADB publications may be found at http://www.adb.org/publications/corrigenda.

Notes:
In this publication, "$" refers to United States dollars.
ADB recognizes "Hongkong" as Hong Kong, China.

Cover design by Francis Manio.

Printed on recycled paper

Contents

Tables and Figures

Foreword

Infrastructure projects help drive economic prosperity and provide a solid base for strong, sustainable, balanced, and inclusive growth and development. Well-prepared infrastructure projects are also key to achieving the Sustainable Development Goals (SDGs), which are now more relevant than ever. As such, SOURCE, an operating platform supported by multilateral development banks to address the financing gap in sustainable infrastructure, is a promising tool for helping countries to design and manage projects aligned with the SDGs and other international agreements.

The coronavirus disease (COVID-19) has highlighted the world's vulnerability to shocks and stresses, and many countries are now even more offtrack in achieving the SDGs than before the pandemic. According to the United Nations Economic and Social Commission for Asia and the Pacific (UNESCAP), the financing gap in 2019 was around $1.5 trillion a year (5% of regional gross domestic product) to fully achieve the SDGs by 2030. To address this gap, governments have implemented recovery programs with the support of the Asian Development Bank and other multilateral and bilateral development partners. Their immediate financial response packages helped to alleviate the public health crisis and its consequences on the real sector.

However, substantially more private sector financing is required to address the enormous needs for economic recovery and mitigate adverse development impacts. History has shown that a crisis can become an opportunity if the disruptions or shocks compel reforms that would otherwise not be considered feasible under normal conditions. The COVID-19 recovery programs offer a unique opportunity to craft a path consonant with green, resilient, inclusive, and sustainable infrastructure development, in line with the SDGs.

Investing in sustainable and quality infrastructure projects is crucial to rebuild smartly and create a more inclusive and equitable world. Major reforms across sectors are still needed to create more resilient societies. As we all witnessed, digital transformation accelerated during the pandemic and has caused major structural shifts to new ways of production, cooperation, and consumption across the world. Similarly, the digitalization of processes related to infrastructure projects and investments provides a key opportunity for driving the agenda of sustainable economic development and collective resilience.

SOURCE is an outstanding example of a digital transformation tool. The platform has been designed through joint efforts among multilateral development banks and enables a systemic transition to the digitalization of infrastructure project development and data collection. SOURCE is a crucial mechanism helping countries to build a pipeline of infrastructure projects aligned with the SDGs as well as to the G20 Principles for Quality Infrastructure Investment. Some 50 countries have expressed interest in using SOURCE. Some have already launched a full integration of SOURCE into their national systems for infrastructure project development, while others are considering and planning pilot projects to test and evaluate the platform's benefits.

This publication will enhance the awareness of SOURCE as a digital platform to support public sector agencies with standardized processes that promote sustainable infrastructure projects. The use of SOURCE can be integral to capacity building for best practices in project management, for building infrastructure databases essential to measuring achievements, and for raising bankability and attracting private sector financing of projects aligned with and facilitating the achievement of SDGs.

Ramesh Subramaniam
Director General and Group Chief
Sectors Group
Asian Development Bank

Acknowledgments

This document was prepared in cooperation with the Asian Development Bank (ADB) and the Sustainable Infrastructure Foundation (SIF) in close coordination between Pedro Nicolau, integration manager at SIF; Thomas Kessler, principal finance specialist (disaster insurance), Finance Sector Office, Sectors Group, ADB, and Kin Wai Chan, markets development advisory specialist, Special Initiatives and Funds, Office of Markets Development and Public–Private Partnership, ADB.

The team was supported by Bruno Carrasco, director general, Climate Change and Sustainable Development Department, ADB; Junkyu Lee, director, Finance Sector Office, Sectors Group, ADB; Adrian Torres, director, special initiatives and funds, Office of Markets Development and Public–Private Partnership, ADB; Trevor Lewis, advisor, Strategy, Policy, and Partnerships Department and head for nonsovereign operations, Strategy, Policy and Partnerships Department, ADB; as well as Christophe Dossarps, chief executive officer of SIF; Cedric van Riel, integration manager at SIF; and Pierre Sarrat, transport engineer, European Investment Bank.

The team benefited from significant ADB internal and external peer-reviewer inputs and feedback by the following:

- Syed Uddin, public–private partnership policy advisor, Ministry of Finance and Public–Private Partnership Development Agency, Uzbekistan
- Jeanine Corvetto, resident advisor, Ministry of Economy and Finance Ecuador, Government Debt and Infrastructure Finance, Office of Technical Assistance, US Department of the Treasury
- Marc Tkach, director, Infrastructure and Integrated Program Management, Millennium Challenge Corporation
- Silvia Brugger, Deutsche Gesellschaft für Internationale Zusammenarbeit (GIZ) coordinator climate governance, EUROCLIMA+ Program
- Christian Déseglise, managing director, Head of Sustainable Finance and Investments, Global Banking and Markets, Hongkong and Shanghai Banking Corporation
- Denis Prouteau, chief investment officer, Private Debt, Natixis Investment Managers
- Christian Yoka, regional director—Eastern Africa, Agence Française de Développement
- Inga Beie, Head of Project Leading Urban Climate Action, GIZ
- Taras Boichuk, head of office, SPILNO Public–Private Partnership Project Management Office, Ministry of Infrastructure of Ukraine
- Feroisa Francisca T. Concordia, director, Public–Private Partnership Center, Philippines

Finally, the document was reviewed and endorsed by all members from the SOURCE Council, including Matthew Jordan Tank, director, Sustainable Infrastructure Policy & Project Preparation at European Bank for Reconstruction and Development; Pedro de Lima, permanent representative of the European Investment Bank Group in the United States, European Investment Bank; Gaston Astesiano, public–private partnerships team leader at Inter-American Development Bank; and Imad Fakhouri, global director, Infrastructure Finance, Public–Private Partnerships and Guarantees Global Practice, World Bank Group.

Abbreviations

ADB	Asian Development Bank
API	application programming interface
COVID-19	coronavirus disease
DMC	developing member country
EBRD	European Bank for Reconstruction and Development
ESG	environmental, social, and governance
FAST-Infra	Finance to Accelerate the Sustainable Transition-Infrastructure
G20	international forum for the governments and central bank governors for 19 countries and the European Union
IMF	International Monetary Fund
IT	information technology
MDB	multilateral development bank
PPP	public–private partnership
QII	quality infrastructure investment
SDG	Sustainable Development Goal
SIF	Sustainable Infrastructure Foundation
UN	United Nations

Executive Summary

The coronavirus disease (COVID-19) pandemic exposed the world's fragile development progress and likewise revealed its capacity for financial resilience when confronted with disasters. This global robustness provides an opportunity to build back better and boost awareness of the urgency and benefit of sustainable development and infrastructure investment. This is especially critical in the face of the Russian invasion of Ukraine and the disruptions in global supply chains that exacerbate the need to boost innovative solutions to address the climate and food emergency.

The worldwide transformation to digital economies is a key driver in the important agenda of sustainable finance. Supporting this transformation is the digital multilateral platform, SOURCE, which helps ensure well-prepared infrastructure projects in governments' digitalization agendas.

SOURCE is a global and scalable information technology platform, hosted by the United Nations, acting as a unique delivery system for the world's best practices in infrastructure project development. Its structured data-based approach is designed to strengthen accountability and transparency and to support informed decisions that meet local, regional, and global standards.

The SOURCE framework is designed to optimize the efficiency of stimulus packages being prepared in response to COVID-19, by ensuring swift deployment of investments while preserving their compliance with objectives such as the Sustainable Development Goals and the Paris Agreement.

SOURCE is led and funded by multilateral development banks through its council composed of the Asian Development Bank (ADB), European Bank for Reconstruction and Development (EBRD), European Investment Bank (EID), Inter-American Development Bank (IADB), and World Bank, with the African Development Bank and Asian Infrastructure Investment Bank as observers.

The Sustainable Infrastructure Foundation (SIF) is the implementing agency managing SOURCE. SIF is registered as a nonprofit foundation established in Switzerland and has official observer status in the G20 Infrastructure Working Group.

SOURCE provides a standard benchmark framework for projects, with a specific approach for each sector and procurement mode. This enables assessment and comparison of the effectiveness of a project or an entire portfolio against benchmarks and standards, such as the Quality Infrastructure Investment principles, the European Union Taxonomy, and the United Nations' Sustainable Development Goals. A systematic digital tagging of environmental, social, and corporate governance assets is intended to create a marketplace attracting investors and mobilizing financing from the private sector. SOURCE's sector-specific sets of questions, covering all stages of the project cycle, strengthen the inputs of public sponsors and help raise awareness among project developers of critical environmental, social, and governance criteria; climate change mitigation and resilience; procurement transparency and openness; inclusiveness; gender equality and equity; employment creation; and financial and fiscal sustainability.

On 18 July 2020, the G20 finance ministers and central bank governors endorsed SOURCE "to enable a systemic transition to the digitalization of infrastructure project development and data collection as part of advancing the work related to the Quality Infrastructure Investment principles." This ambition reflects the coordinated efforts of multilateral development banks to exercise shared leadership for the benefit of their members.

Introduction

Background

The United Nations Sustainable Development Goals. Infrastructure investment is a key component of the United Nations (UN) 2030 Agenda for Sustainable Development[1] and is recognized as a crucial driver of economic development. This Agenda for Sustainable Development includes 17 Sustainable Development Goals (SDGs), each with its respective targets and indicators and plan for implementation. Infrastructure is an explicit goal (SDG 9) and an implicit means to implement and achieve other SDGs. Since the quality, quantity, and accessibility of economic and social infrastructure in developing countries lag the advanced economies, scaling up infrastructure investment is a key pillar in many national development strategies.

Infrastructure gap. For most emerging and developing countries, the major obstacle in mobilizing and channeling private capital in infrastructure is the lack of a pipeline of well-prepared, bankable, and investable projects. The incentives are low for potential project sponsors to commit to the high development costs of bid proposals, due to project implementation uncertainties. The inability of sovereign and sub-sovereign agencies to prepare a consistent supply of structured projects offering a standardized risk allocation scheme and satisfactory investment return remains a major bottleneck. This is because projects suffer from poor development and unclear legal and regulatory frameworks, leading to delays and a lack of transparency in the bidding process. Bridging large infrastructure gaps will therefore require greater capacity in countries to identify, structure, and procure projects that meet international standards (including governance, social, and environmental). The multilateral development banks have all acknowledged this need and are thus committed to private sector development and to increase private sector operations significantly toward 2030.

Need for quality infrastructure. In 2019, Japan assumed the G20 Presidency,[2] and the major infrastructure outcome was the endorsement of the G20 Principles for Quality Infrastructure Investment. The preamble to the principles notes a renewed emphasis on quality infrastructure investment that will build on past G20 presidency efforts to mobilize financing from various sources. This includes private sector and institutional sources (including multilateral development banks) in particular to help close the infrastructure gap, developing infrastructure as an asset class, and maximizing the positive impacts of infrastructure investment according to country conditions.[3] The G20 Principles include the following to promote quality infrastructure investment:

(i)   Maximize the positive impact of infrastructure to achieve sustainable growth and development.
(ii)   Raise economic efficiency in view of the life-cycle cost.
(iii)   Integrate environmental considerations in infrastructure investments.

---

[1]   United Nations. 2015. Transforming Our World: The 2030 Agenda for Sustainable Development. https://sustainabledevelopment.un.org/post2015/transformingourworld/publication.
[2]   G20. 2019. https://g20.org/en/about/Pages/whatis.aspx.
[3]   G20. 2019. https://www.mof.go.jp/english/international_policy/convention/g20/annex6_1.pdf.

(iv) Build resilience against disasters and other risks.
(v)   Integrate social considerations in infrastructure investment.
(vi) Strengthen infrastructure governance.

Why SOURCE?

SOURCE is a critical tool that helps to answer the following questions in detail:

- How can an effective project management system enhance coordination among the many stakeholders during the complex lifecycle of an infrastructure project?
- How will it help to have a standardized workflow process to foster accountability and transparency and to allow checks against targeted performance indicators?
- How can the capacity of implementing agencies be strengthened to provide them an information management system that gives access to best practice procedures and to all relevant data available for each stakeholder at each step of the process?
- How can worldwide knowledge and experience of development organizations be shared and effectively embedded for better informed decision-making?
- How can a structured infrastructure database set a global standard for infrastructure as an asset class promoting environmental, social, and governance project investments and mobilizing private sector financing?
- How can governments effectively monetize the available structured data and more effectively manage procurement processes?
- How can a marketplace be built and achieve a common understanding between public and private investors and coordination between institutional investors, multilateral development banks, and governments?
- How can the process facilitate access to funds to support project development and promote public–private partnerships?
- How relevant is SOURCE to enhance resilience against infectious diseases of critical infrastructure and to monitor enhancements achieved in this area?
- How will such an online solution platform support administration in "work-from-home" settings under imposed confinements and ensure a certain project continuity?

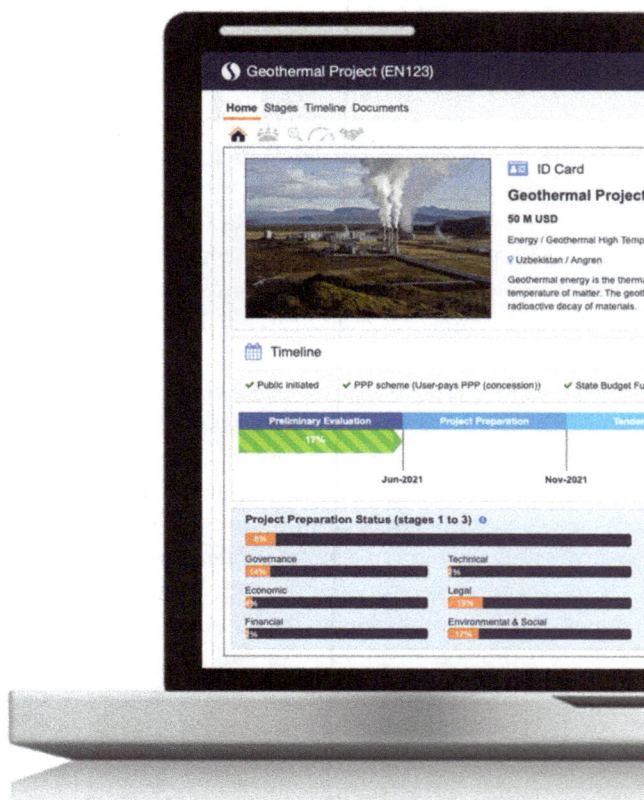

Multilateral Development Banks Working as a System: A Success Story

Phase 1: Development of a Digital Knowledge Dissemination Tool

The Asian Development Bank (ADB) originated SOURCE as the National Infrastructure Information System under a regional technical assistance project[4] approved in 2009 for the benefit of the organization's developing member countries. In 2013, the platform was renamed the International Infrastructure Support System and its management transferred to the Sustainable Infrastructure Foundation (SIF), a not-for-profit organization registered in Switzerland, to make it an international tool to be used widely by all interested parties and for multilateral development banks to collaborate more easily.

ADB's technical assistance subsequently supported the upgrading of the platform and the development of the original infrastructure templates until 2014. The platform was eventually launched globally in 2016 and finally renamed SOURCE in 2017.

In this form, SOURCE provided public sponsors with a collaborative and secure workspace, based on sub-sectorial templates built on a compendium of good practices jointly designed with the private sector. SOURCE's templates covered all aspects of the project, i.e., governance, technical, economic, legal, financial, environmental, and social. The templates come from and are updated with inputs from leading private sector investors/sponsors/operators/financiers and advisors. This effectively takes best practice and knowledge from the private sector and distills it into usage/guidance that can be used from the outset of a government team's efforts to start preparing a project. So far, more than 7,000 comments have been provided by the private sector, making SOURCE a standardized dataset creating the international common language between the public and private sectors.

During an initial outreach phase during 2016–2019, the platform benefited more than 3,200 users across 59 countries and supported the development of close to 400 infrastructure projects worldwide.

Since 2014, SOURCE has been guided and funded by multilateral development banks and Project Preparation Facilities. In 2021, the cumulative contributions from multilateral development banks were around $5.7 million (ADB $1.4 million).

---

[4]   RETA 7379: Establishment of e-Systems in Support of Infrastructure Finance in Asia.

Phase 2: Evolution into a Fully Operational Tool and Start of the Integration Strategy

Since 2019, the strategic and financial management of SOURCE has been under the supervision of the SOURCE Council composed of representatives from major multilateral development banks including ADB, the European Bank for Reconstruction and Development (EBRD), the European Investment Bank, Inter-American Development Bank, International Finance Cooperation, and the World Bank Group. The African Development Bank and the Asian Infrastructure Investment Bank are observers. The SOURCE Council enjoys the following powers:

- define and direct the operations related to the development, maintenance, and management of the SOURCE platform;
- oversee the activities of SIF in accordance with the agreed revenue model for the distribution of the SOURCE platform; and
- advise SIF to strengthen SOURCE's performance and scalability.

In March 2019, the SOURCE Council endorsed a new strategy for deploying SOURCE in its member economies via an integration process.

This came after a stock-taking exercise that concluded that the initial approach followed was not meeting SOURCE's transformational agenda. Irrespective of the usefulness and practicality of the platform, an institutional framework is required to ensure efficient and sustainable implementation and adoption by government agencies. All SIF trainings conducted in the first phase of deployment of the platform have been central to the dissemination and promotion of the tool, and allowed professional practitioners to test the software and provide their feedback. But punctual trainings eventually brought little change at the organizational level, which is where SOURCE brings its full benefits.

The integration strategy was therefore developed, whereby SIF customizes SOURCE to the local regulatory framework and enables its adoption by all relevant agencies through a vast campaign of training. Following the successful integration, SIF guarantees the continuity of services and provides technical support to user countries at no cost to the government. The process of integrating SOURCE encompasses the following workstreams, in accordance with the process described in Figure 1:

- **Policy stream**—which entails a review of the primary and secondary legislation potentially conflicting with the implementation of SOURCE platform for infrastructure project delivery and the development of recommendations to the regulatory approach that will aim to eliminate conflicts and implement adjustments to the platform itself.
- **Process stream**—which identifies and maps out current processes, taxonomy, and governance to appropriately customize the SOURCE platform to the needs and framework of the country.
- **IT tools stream**—which focuses on technical integration of the SOURCE platform and development of technical adjustments required to ensure interoperability with the existing IT tools and databases, and its testing and user training.

To address the above needs and enable efficient adoption by governments, SOURCE has evolved from an online knowledge dissemination vehicle to a complete information technology solution (Figure 2).

Figure 1: Integration Process

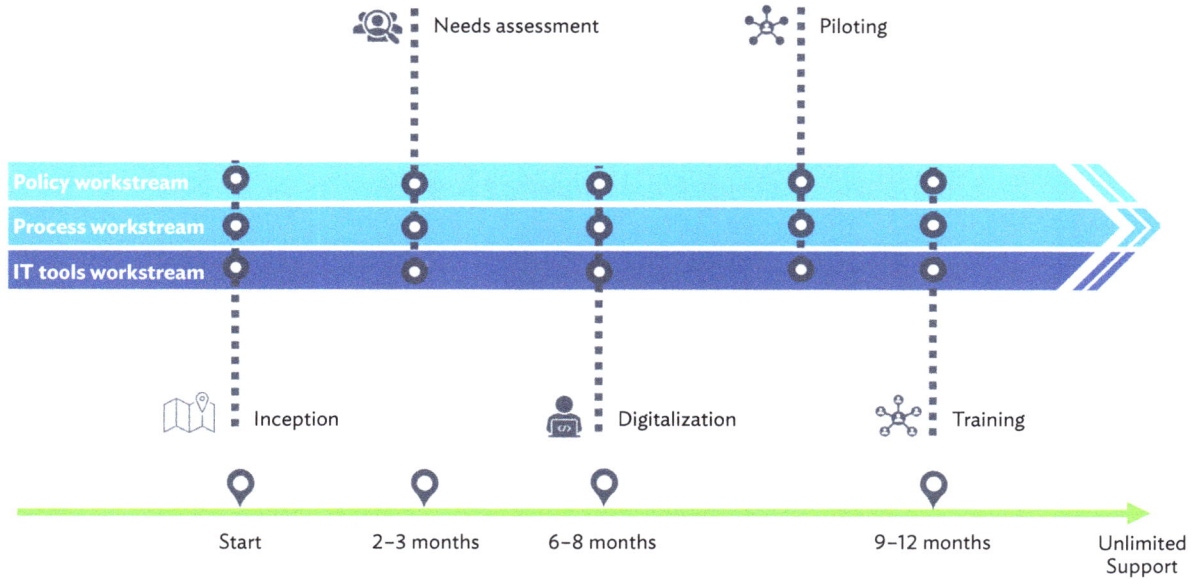

Needs assessment

Piloting

Policy workstream

Process workstream

IT tools workstream

Inception

Digitalization

Training

| Start | 2–3 months | 6–8 months | 9–12 months | Unlimited Support |

IT = information technology.
Source: Sustainable Infrastructure Foundation, 2023.

Figure 2: Key Milestones of SOURCE's Development

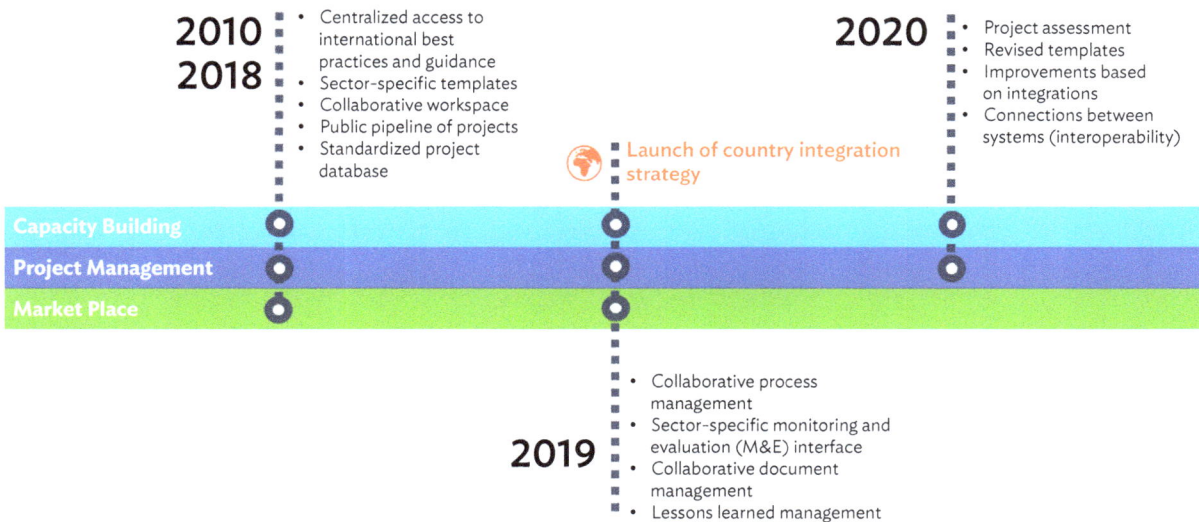

2010
2018
- Centralized access to international best practices and guidance
- Sector-specific templates
- Collaborative workspace
- Public pipeline of projects
- Standardized project database

2020
- Project assessment
- Revised templates
- Improvements based on integrations
- Connections between systems (interoperability)

Launch of country integration strategy

Capacity Building

Project Management

Market Place

2019
- Collaborative process management
- Sector-specific monitoring and evaluation (M&E) interface
- Collaborative document management
- Lessons learned management

Source: Sustainable Infrastructure Foundation, 2021.

Phase 3: Scaling Up Integration toward Financial Sustainability

The officialization of the executive role of multilateral development banks in SOURCE's governance through the creation of the SOURCE Council proved central to the scaling up of SOURCE's dissemination in global events, recognition of its central role by the G20 Infrastructure Working Group, the Organisation for Economic Co-operation and Development, UN agencies, and adoption by governments.

The Summit on the Financing of African Economies in Paris in May 2021, for instance, led to the endorsement of SOURCE and its deployment in Africa by more than 50 heads of state. Appendix 1 lists major references to SOURCE in international publications.

In June 2020, the SOURCE Council approved a revenue model based on a plan to complete the integration of SOURCE in 40 countries within a 5-year period, funded by donors. This revenue model aims to leverage the massive surge in the number of users and information collected on the platform to develop diverse revenue streams that will gradually replace the current financial support from multilateral development banks and ensure SOURCE's long-term financial sustainability. While SOURCE will continue offering business intelligence features at no cost for public sector users to provide insights, analytics, benchmarks, and trends on their pipeline and their region to make better informed decisions, revenue will be generated by monetizing anonymized/aggregated business reports for the private sector.[5] In the 5-year ramp-up period so far, SIF and multilateral development banks have already started to coordinate and raise donor funding to provide part or all of the required financial support to facilitate the required country integrations, while the revenues generated will cover future integrations with or without donors.

SOURCE's global demand from governments around the world keeps growing (Figure 3) and the SOURCE Council's key priorities are identifying solutions to raise more funds to meet the demand coming from their members and continuing to support and ensure that the SIF can successfully implement the revenue model for financial sustainability.

SOURCE can play a key role in helping countries overcome the COVID-19 crisis in global supply chain consequences by supporting the digitalization agenda, reinforcing coordination, and reducing competitiveness bottlenecks.

---

[5] The quality of reports is only as good as the quality of data – this is a challenge faced by all multilateral development banks, international organizations, and private system databases. For this, several features to enable data verification by third parties have been integrated into SOURCE, and SIF is continuously improving platform capability to conduct automatic quality assessment.

Figure 3: Outreach Status of SOURCE in 2023

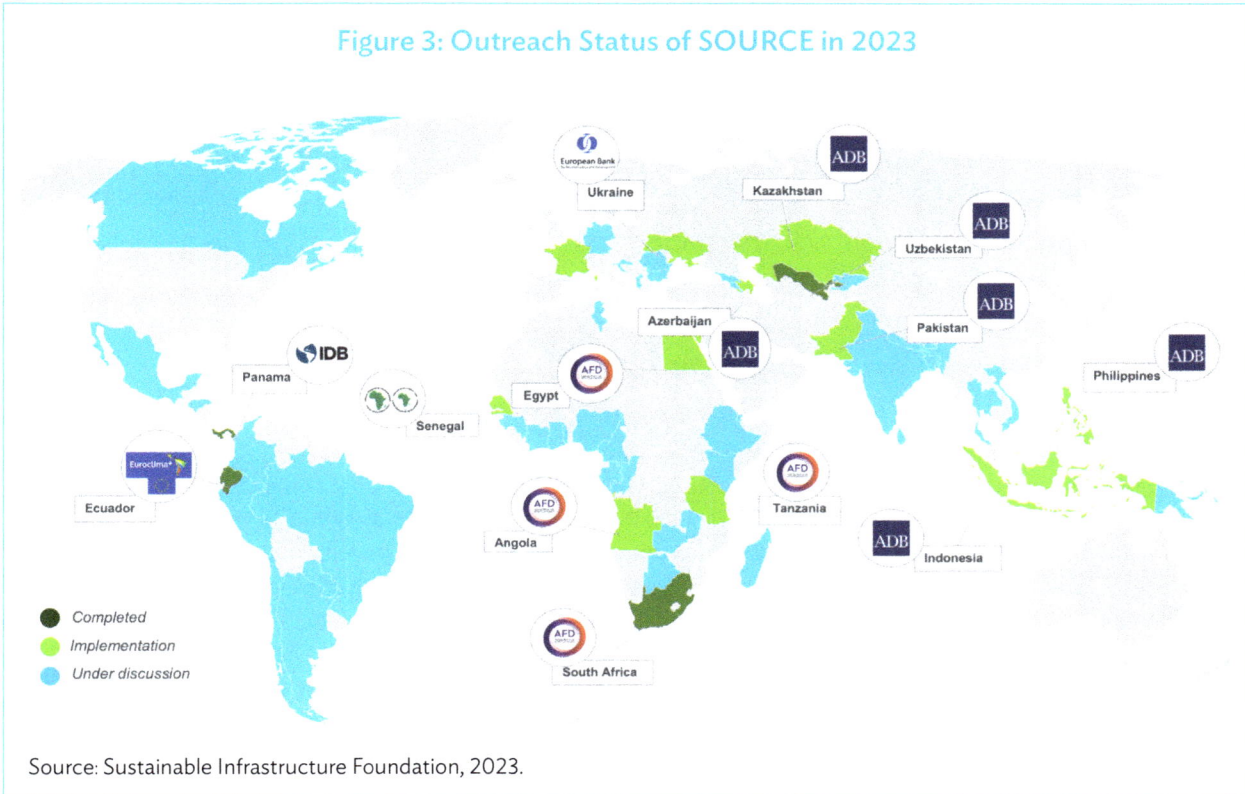

Source: Sustainable Infrastructure Foundation, 2023.

SOURCE's Key Value Propositions

Operationalizing Multilateral Development Bank Knowledge Products

SOURCE templates break down the whole project cycle into well-defined individual steps. This enables the platform to deliver international tools, guidance, and best practices to the right audience at the right juncture in decision-making. It allows a wider array of projects than the one multilateral development banks are supporting. SOURCE already embeds references to World Bank documents and guidance, to the Association of Project Managers Group PPP Guide, the International Finance Corporation performance standards, and many others (Table 1). Further, multilateral development banks' screening and scoring tools can be embedded into SOURCE to drive the UN Sustainability Agenda and the banks' key operational goals. It can disseminate these methodologies to all the banks' members and positively influence decision-making in early project stages.

The MDB Infrastructure Cooperation Platform notes in the guidance note on project development[6] that the "SOURCE platform should be rather considered as an enabling platform to deliver scale and consistency to the entire workstream presented in the document. Through its structured approach to project data and the growing adoption of the platform by national government agencies and multilateral development banks, SOURCE would also enable the collection of standardised project data globally and the usage of MDB tools." In the same document, SOURCE occupies a unique position across the full project cycle, as per the table reproduced in Figure 4.

---

6   EBRD and World Bank Group. 2019. MDB Infrastructure Cooperation Platform: Project Preparation Workstream—Phase II Reference Note on Project Preparation across the Full Project Cycle. Washington, DC.

Figure 4: Overview of MDB Support across the Infrastructure Investment Cycle

		Project-Related Tools and Instruments Used by MDBs	Constraint Addressed	Linkages to Quality
S O U R C E	**Enabling Environment and Institutional Capacity Building**	• Country PPP Readiness Diagnostic • PPP Practitioners Certification Course Procuring Infrastructure PPPs • Framework for Disclosure in PPPs • Public Investment Management Assessment (PIMA) • Procuring Infrastructure PPPs	The need for sound institutional, legal, and regulatory environment to attract long-term private sector participation and investment in infrastructure.	Principle 6: Strengthening Infrastructure Governance
	Infrastructure Planning and Prioritization	• PPP Screening Tools • Infrastructure Prioritization Framework	The need for governments to prioritize projects, public and private, within limited resources.	Principle 2: Raising Economic Efficiency in View of Life-Cycle Cost Principle 6: Strengthening Infrastructure Governance
	Project Design and Appraisal (Pre-feasibility/Feasibility) and Structuring	• PPP Fiscal Risk Assessment (P-FRAM) • Project Readiness Assessment • Guidance on Standard PPP Contract Provisions • Sustainable Infrastructure Indicators	The lack of projects and programs that are commercially viable, affordable, sustainable, and which offer VfM to taxpayers and users.	Principle 3: Integrating Environmental Considerations Principle 4: Building Resilience against Natural Disasters and Other Risks Principle 5: Integrating Social Considerations Principle 6: Strengthening Infrastructure Governance
	Bid Design and Procurement through Commercial/Financial Close	• Guidance Note on Procurement for Quality Infrastructure • Guidelines for Managing Unsolicited Proposals	The inability of governments to execute the transaction process effectively, allocate risk efficiently, and get a fair deal.	Principle 6: Strengthening Infrastructure Governance
	Implementation and Monitoring	• PPP Contract Management tool	The difficulty of governments to monitor private sector obligations in the form of contractual KPIs/performance standards, eroding potential VfM.	Principle 2: Raising Economic Efficiency in View of Life-Cycle Cost Principle 5: Integrating Social Considerations

(PPP REFERENCE GUIDE)

KPI = key performance indicator, MDB = multilateral development bank, PPP = public–private partnership, VfM = value for money.

Source: EBRD and World Bank. 2019. Guidance Note on Project Preparation.

Table 1: Tools, Guidance, and Best Practices Embedded into SOURCE

Tools, Guidance, and Best Practices	Type of Integration in SOURCE			
	Related Questions/ Data Points in Templates	Dedicated Questions/ Data Points in Templates	Dedicated Tooltip(s) with External Link	Dedicated Webpart/ Assessment Methodology
UN Sustainable Development Goals	●	●	●	●
G20 Principles for Quality Infrastructure Investment	●	■		■
World Bank Guidance on PPP Contractual Provisions	●		●	
World Bank Policy Guidelines for Managing Unsolicited Proposals	●		●	
World Bank PPPI Database and Benchmarking Reports	●			
World Bank–International Monetary Fund PPP Fiscal Risk Assessment Model Tool	●	●	●	
International Finance Corporation Performance Standards	●		●	
PPP Guide (CP3P Certification Scheme, Association of Project Managers Group)	●			
Organisation for Economic Co-operation and Development Principles for Public Governance of Public–Private Partnerships	●		●	
UNCITRAL Legislative Guide on Public–Private Partnerships	●		●	
Inter-American Development Bank Sustainable Infrastructure Criteria	●			
Global Infrastructure Hub PPP Risk Allocation Matrix	●			●
Equator Principles	●	●	●	
International Labour Organization's Decent Work Agenda	●		●	
Paris Agreement (Nationally Determined Contributions)	●	●	●	
ADB's GRIS Methodology	●			●
FAST-Infra Label	●			●
UNECE PPP for SDG				●
G20 QII Indicators				■

● Available in 2021   ■ To be integrated into the next version of templates (2023)

ADB = Asian Development Bank, CP3P = Certified Public–Private Partnerships Professional, FAST-Infra = Finance to Accelerate the Sustainable Transition – Infrastructure, G20 = Group of Twenty, GRIS = green, resilient, inclusive, and sustainable, PPP = public–private partnership, QII = Quality Infrastructure Investment, SDG = Sustainable Development Goal, UN = United Nations, UNECE = United Nations Economic Commission for Europe, UNCITRAL = United Nations Commission on International Trade Law.

Source: Sustainable Infrastructure Foundation, 2023.

Enabling Sustainable Development Goal Measurement

The standardized project development templates on SOURCE cover both traditional procurement methods and public–private partnerships. From project definition to operation and maintenance, SOURCE allows public agencies to define clear SDG targets early, during development, that can be integrated and monitored across the lifecycle. SOURCE, therefore, empowers government agencies to set up quantified targets, defining clear expected positive impacts related to the SDGs, going further than simple declaration of good intentions. These targets and positive impacts can then be monitored during construction and operation and maintenance to make sure projects comply with set targets. Automatic quality assessment of project documentation may be done across the stages by the "consistency checks" function in SOURCE. Indeed, a red flag appears when an SDG is targeted but not translated into quantified targets as part of the expected positive impacts identified during the project in the appraisal stage. An SDG impact dashboard (Figure 5) is also available at the portfolio level in SOURCE. It provides a visual representation of the key indicators of the portfolio of projects managed in SOURCE against each relevant SDG.

> *"The value and experience that has been accumulated by SOURCE and the SIF team is absolutely critical to the success of the platform. SOURCE can play a critical role in monitoring data. The standardized presentation of the project which is offered by SOURCE and the customization which covers the specificities of the different countries makes the choice of SOURCE, a very logical one to be at the beginning of the end-to-end platform. This will allow for the data which is hosted by SOURCE to flow more freely towards potential funders."*
>
> **Christian Deseglise**
> Head of Sustainable Finance and Investments
> HSBC | FAST-Infra

Moreover, SOURCE can also assess projects regarding its gender equality targets, according to the SDG 5 (Achieve gender equality and empower all women and girls). The European Bank for Reconstruction and Development (EBRD) and SIF contracted a gender consultant with the specific objective of developing and providing the gender- and inclusion-related actions as part of the SOURCE project development tool.

Providing a Global Information Technology Solution that Ongoing Initiatives Need

In addition to be the digital dissemination platform for international standards, SOURCE is a global information technology (IT) solution that can host and deliver existing tools. For example, SIF embedded the PPP Risk Allocation Tool into SOURCE with the support of the Global Infrastructure Hub to enable informing of project developers about the allocation of risks—as between the public and private partners in typical PPP transactions for 18 different types of projects—by answering the questions in the SOURCE templates. Delivery challenges from the Global Delivery Initiative were also embedded into SOURCE. Based on the project's geographical region, sector SOURCE provides a list of the most recurring delivery challenges to alert the project team and help them in risk assessment.

SOURCE can also be part of IT systems designed by ongoing initiatives to feed complementary platforms with standardized data. The Finance to Accelerate the Sustainable Transition-Infrastructure (FAST-Infra)[7] initiative identified SOURCE as the central component of its end-to-end platform technology platform. The projects initiated by public entities will therefore be originated from SOURCE, which will also enable the auto-assessment

---

7   HSBC. 2021, https://www.sustainablefinance.hsbc.com/sustainable-infrastructure/fast-infra-a-public-private-initiative.

Figure 5: SDG Impact Dashboard in SOURCE

The 17 Sustainable Development Goals (SDGs) were adopted by all United Nations Member States in 2015, as part of the 2030 Agenda for **Sustainable Development**. It is a call to urgent action by all countries within the framework of a global partnership. Link: https://sdgs.un.org/goals. The graphics below present the sum of the targeted positive impacts related to SDGs already defined in the Stage 2 of your project pipeline.

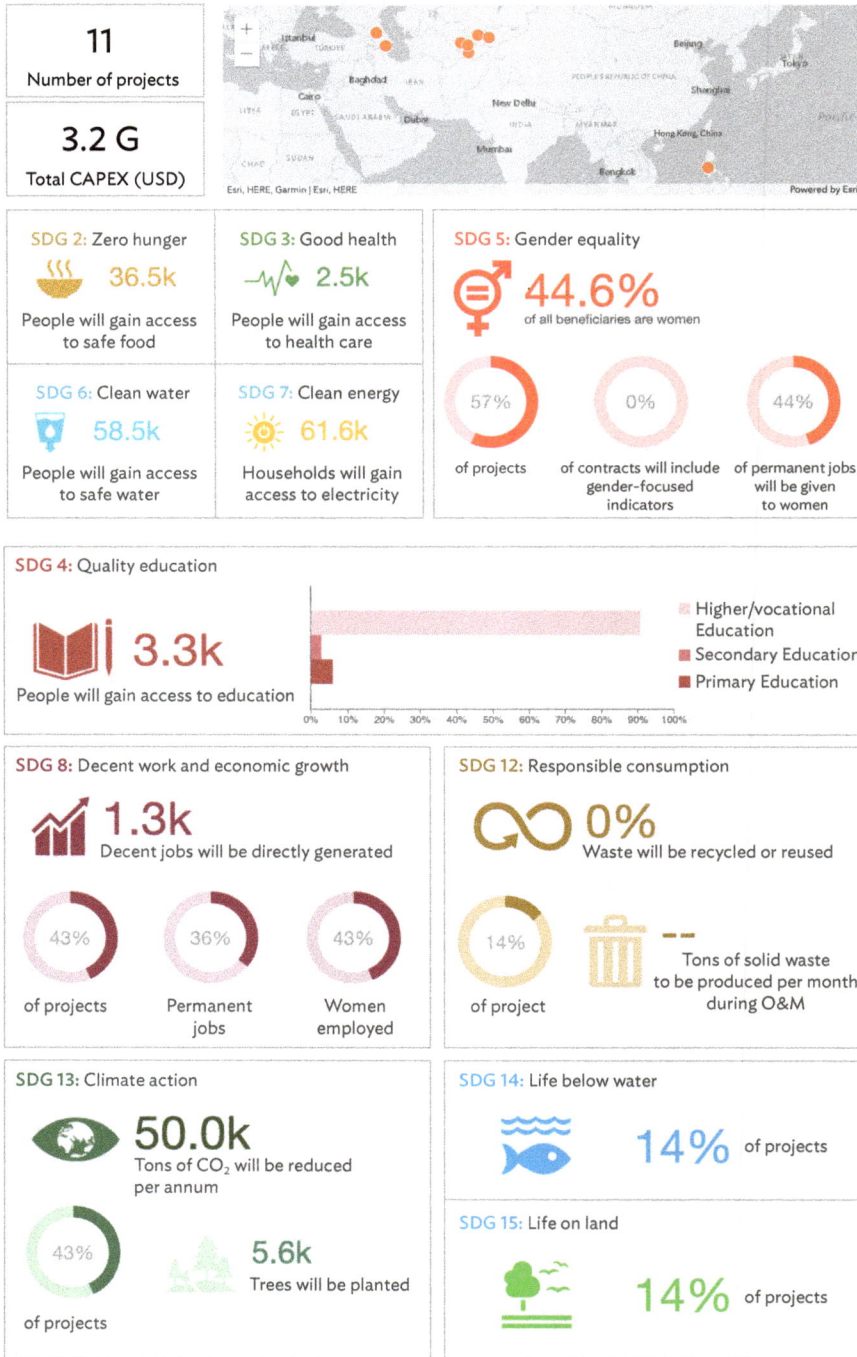

11
Number of projects

3.2 G
Total CAPEX (USD)

SDG 2: Zero hunger
36.5k
People will gain access to safe food

SDG 3: Good health
2.5k
People will gain access to health care

SDG 5: Gender equality
44.6%
of all beneficiaries are women

57% of projects
0% of contracts will include gender-focused indicators
44% of permanent jobs will be given to women

SDG 6: Clean water
58.5k
People will gain access to safe water

SDG 7: Clean energy
61.6k
Households will gain access to electricity

SDG 4: Quality education
3.3k
People will gain access to education

Higher/vocational Education
Secondary Education
Primary Education

SDG 8: Decent work and economic growth
1.3k
Decent jobs will be directly generated

43% of projects
36% Permanent jobs
43% Women employed

SDG 12: Responsible consumption
0%
Waste will be recycled or reused

14% of project
Tons of solid waste to be produced per month during O&M

SDG 13: Climate action
50.0k
Tons of CO_2 will be reduced per annum

43% of projects
5.6k
Trees will be planted

SDG 14: Life below water
14% of projects

SDG 15: Life on land
14% of projects

Source: Sustainable Infrastructure Foundation, 2022.

of their project documentation against the FAST-Infra label before transferring the relevant data to the FAST data platform for the deal origination.

SOURCE provides a comprehensive project management system, which reinforces high-quality gatekeeping adapted to country frameworks, and embeds a compendium of best practices and guidance references from the international community and the private sector. This combination of services, and the positioning of the platform across the full project cycle, makes SOURCE a unique offering from multilateral development banks and international organizations' perspectives. It is also a unique offering among commercial solutions, which focus on process management functionalities (MS Project, Primavera, etc.) and information management (Dropbox, Ansarada, etc.) and/or provide specific services for well-identified revenue-generating segments of the project cycle (e.g., BIM, data room solutions). Figure 6 summarizes the full array of SOURCE's value propositions with key representative solutions providing similar services.

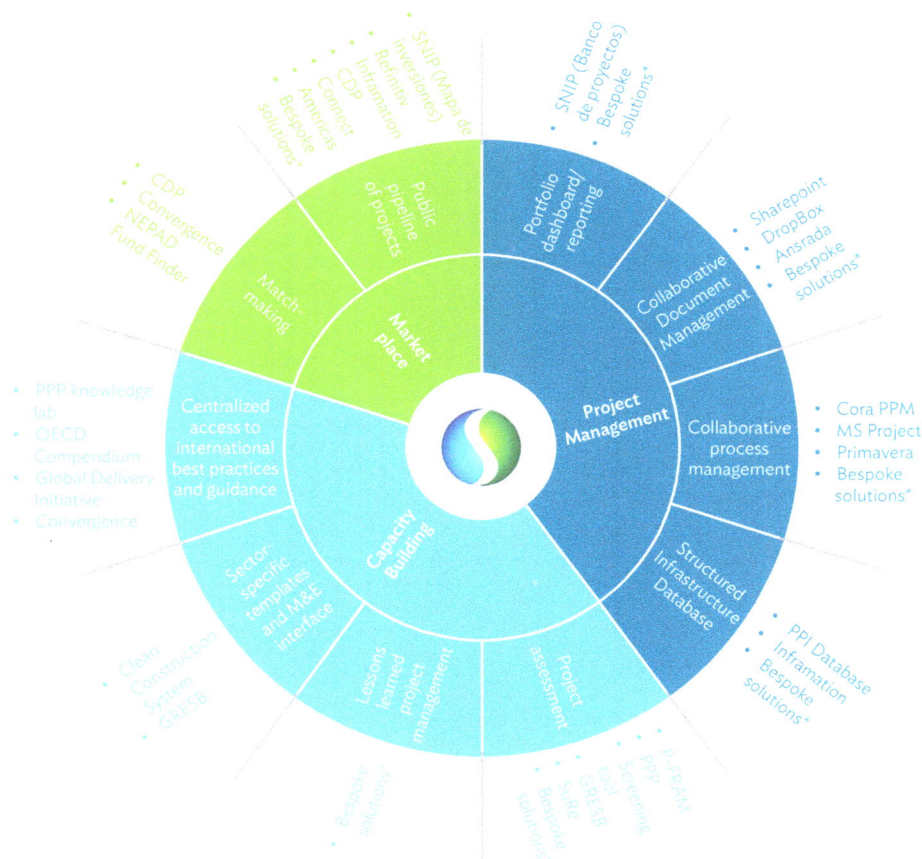

Figure 6: SOURCE Value Proposition and Market Positioning

CDP = Carbon Disclosure Project, GRESB = Global Real Estate Sustainability Benchmark, M&E = monitoring and evaluation, NEPAD = New Partnership for Africa's Development, OECD = Organisation for Economic Co-operation and Development, P-FRAM = PPP Fiscal Risk Assessment, PPI Database = Private Participation in Infrastructure, SNIP = Sistema Nacional de Inversión Pública, PPP = public–private partnership, SuRe = Standard for Sustainable and Resilient Infrastructure.

* Bespoke solutions refer to new IT tools that are developed for a specific user or government, by opposition to "off-the-shelf" solutions.

Source: Sustainable Infrastructure Foundation, 2022.

The multilateral development bank-led governance of SOURCE is also appropriate to address governments' confidentiality and sovereignty concerns[8] similar to the Global Emerging Markets Risk Database Consortium or other solutions supported by the multilateral development banks. These include, for example, SNIP[9] in Latin America, but hosted on countries' servers, incurring a maintenance and development cost), or bespoke solutions developed for governments (e.g., the MIS[10] in Indonesia, developed with the World Bank's support but also hosted on the country's servers incurring a maintenance and development cost). These bespoke solutions meet some of the same needs as SOURCE and provide more flexibility to governments in customization and reporting. For this purpose, SOURCE is being implemented in these governments where they use such bespoke solutions as a complementary solution (particularly to facilitate data collection and sharing among all users and IT tools), based on a robust application programming interface (API)[11] strategy, that aims to remove any overlap or need for double-data entry and reduces the challenges of change management as much as possible.

Supporting Infrastructure as an Asset Class

The multilateral SOURCE platform is led and funded by multilateral development banks to support the development of well-prepared infrastructure projects and government digitalization agenda. It provides a comprehensive project management system, which reinforces a high-quality gatekeeping process that can be adapted to regional, national, and subnational regulatory frameworks and embeds a compendium of best practices and guidance references from both the international community and the private sector.

Through sector-specific templates and adaptive timelines, SOURCE allows the capturing and structure of up to 2,000 data points for each project, from early definition to the infrastructure's end of life (see Figure 7).

Information thus collected and curated on SOURCE via a host of sanitizing and decision-making tools may be disclosed by public sponsors in a consistent and lawful manner, at the relevant stages of the projects. SOURCE thereby establishes a *de facto* **universal data standard** for infrastructure projects, facilitating the private sector's origination and due diligence processes (e.g., project readiness; risk allocation; revenue profile; and environmental, social, and governance). This combination of services, and the positioning of the platform across the full project cycle, makes SOURCE a unique offering, based on the following:

- **The multilateral governance of SOURCE**, embodied by the consortium of multilateral development banks at the SOURCE Council,[12] which guarantees the independence and sustainability of the platform.
- **The strong involvement of the institutional and private sectors in the elaboration of SOURCE's data framework**, ensuring all information important for decisions to bid are covered. A total of more than 7,000 comments were received from contributors from the contractors, investors, lenders, and advisors listed in Table 2, ensuring coverage of all relevant decisions for bids.

---

[8]   The United Nations International Computing Centre is the cloud provider of the SOURCE application and users' data. As a United Nations entity, the center operates under the same privileges and immunities as its clients and partner organizations. SOURCE servers are therefore extra-territorial, meaning they are not under the jurisdiction of any national authority. SOURCE data and systems are therefore protected from any national interest or intrusion.

[9]   Sistema Nacional de Inversión Pública.

[10]   Management Information System is a bespoke solution funded by the World Bank developed for the planning agency of Indonesia (BAPPENAS) to manage public–private partnership projects.

[11]   An API is a way to programmatically interact with a separate software component or resource.

[12]   As of 2023, the SOURCE Council is composed of ADB, the European Bank for Reconstruction and Development, the European Investment Bank, the Inter-American Development Bank, the International Finance Corporation, and the World Bank. The African Development Bank and the Asian Infrastructure Investment Bank are as observers.

Figure 7: Typical Distribution of Data Points Collected in SOURCE for One Project

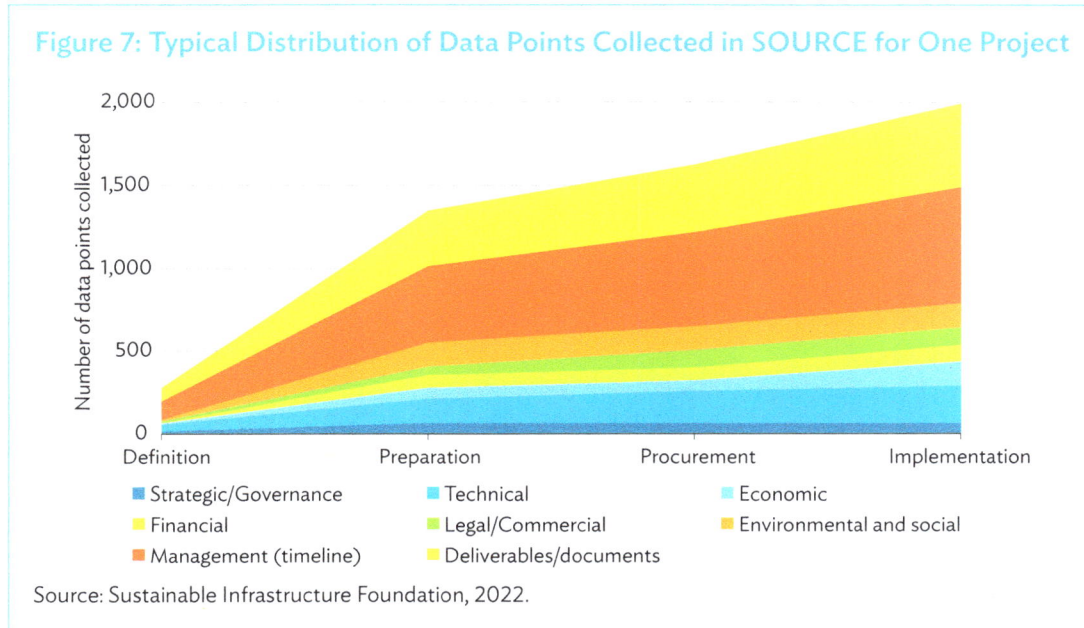

Y-axis: Number of data points collected (0, 500, 1,000, 1,500, 2,000)

X-axis: Definition, Preparation, Procurement, Implementation

Legend:
- ■ Strategic/Governance
- ■ Technical
- ■ Economic
- ■ Financial
- ■ Legal/Commercial
- ■ Environmental and social
- ■ Management (timeline)
- ■ Deliverables/documents

Source: Sustainable Infrastructure Foundation, 2022.

Table 2: Private Contributors to SOURCE Templates

• AECON	• Citibank	• Mott MacDonald
• AIMCo	• CPP Investment Board	• Natixis
• Allianz	• Egis	• OMERS
• APG	• Eiffage	• OPTrust
• Association of Project Managers Group International	• ENGIE	• OTPP
• Ardian	• ERM	• PGGM
• Atkins	• Ferrovial	• Pinsent Masons
• Autodesk	• Generali	• Prudential
• Aviva	• HSBC	• PWC
• Bechtel	• KPMG	• SAP
• BNP Paribas	• Lazard	• SMBC
• Bombardier	• Macquarie	• SNC Lavalin
• Bouygues	• Marsh	• Swiss Re
• CalPERS	• Meridiam	• Veolia
• CDPQ	• Microsoft	• Vinci

Source: Sustainable Infrastructure Foundation, 2022.

The relevance of SOURCE to facilitate the involvement of the private sector in the financing of infrastructure has been confirmed and highlighted by several representative groups of investors, including the Investor Leadership Network[13] and Fast-Infra[14] representing several trillions of dollars.

> *"SOURCE is not only making Infrastructure private debt investment immediately achievable, it also becomes an efficient long-term partner in our quest for welcoming additional developing countries in our investment scope."*
>
> **Denis Protea**
> Chief Investment Officer Private Debt
> Natixis | Investor Leadership Network

Relevance of SOURCE Post COVID-19

As an online project management system and global knowledge dissemination vehicle, SOURCE provides the following specific supports to strengthen the resilience of infrastructure and administration during global crises:

- **Leveraging global lessons learned:** In critical situations, such as during COVID-19, infrastructure sectors must remain resilient, including energy, health care, water and wastewater systems, etc.). One lesson from the crisis is that there is a need to redefine the global approach (toward risk modeling, demand forecasts, sensitivity testing, etc.) to better design and strengthen the resilience of these key sectors. SOURCE has integrated such elements in its sector-specific templates (focusing first on these key sectors) and progressively integrating into SOURCE the guidelines and knowledge products that will emerge from this shared experience to scale up their dissemination.

- **Supporting continuity of operations in crisis:** The global lockdowns in most countries during COVID will be a reminder of the importance and benefits of efficient online solutions, such as SOURCE, that facilitate collaboration and management of information remotely, without need for special settings or access to hard copies of key documents.

- **Assessing the efficiency and global impact of measures:** SOURCE could be adapted and used to help harvest data from different initiatives and assess their effectiveness and socioeconomic impacts (e.g., on the number of permanent jobs created, fiscal revenues, positive externalities, etc.) in a transparent and standardized manner. With investor support, leveraging private investment could also be monitored and facilitated.

- **Scaling up infrastructure project structuring while addressing investors and private sector concern:** SOURCE can help countries to scale up its project structuring processes and build a pipeline that is in line with private investors' needs. SOURCE can help countries in need of faster recovery and reconstruction of essential infrastructure, an example is Ukraine, that by implementing SOURCE will be able to build a consistency project pipeline.

---

[13] The Investor Leadership Network was launched at the 2018 G7 to facilitate and accelerate collaboration by leading global investors with over $9 trillion in assets under management on key issues related to sustainability and long-term growth. The Investor Leadership Network Sustainable Infrastructure Fellowship Program is an executive training program for senior-level public–sector officials in emerging markets' governments, ministries, or agencies responsible for infrastructure planning, development, and management. See https://www.investorleadershipnetwork.org/en/.

[14] HSBC. 2021. Fast-Infra: A Public–Private Initiative. https://www.sustainablefinance.hsbc.com/sustainable-infrastructure/fast-infra-a-public-private-initiative.

Conclusion

SOURCE is a unique and structured approach for infrastructure project development data across the project cycle. Through structured data points, government proponents can aspire to achieve their project's compliance with relevant policy, global development themes, private sector financing priorities, sustainability checklists, quality infrastructure principles, and others. In addition, with a structured approach, bilateral funding agencies, multilateral development banks, and even the private sector can manage their operations and filter more effectively for their specific policy priorities and investment requirements, which in the end helps drive change and quality.

> *"The multilateral platform, SOURCE, enables a systemic transition to the digitalization of infrastructure project preparation and data collection as part of advancing the work related to the QII principles."*
>
> **G20 Finance Ministers and Central Bank Governors**
> G20 Riyadh InfraTech Agenda

SOURCE allows the operationalization of the millions of dollars spent on developing best-practice knowledge products, guidance, and tools, by integrating them into the SOURCE workspace. This integration can help advocacy, knowledge, and funding to be linked, but it all requires structured data. The collection of structured data enables SOURCE to generate reports, benchmarks, and analytics at the project, country program, or global levels. It does this by automatically processing, assessing, and comparing relevant indicators of governance, socioeconomic impacts, financial sustainability, resilience, environmental, and social impacts.

The SOURCE templates are regularly enhanced, now having received more than 7,000 comments and inputs from international organizations as well as some of the world's leading developers, investors, financiers, legal advisers, and operators of infrastructure projects. SOURCE templates are categorized for eight stages of project development (from project definition to operations) and six general aspects of project development (governance, technical, economic, legal, environmental and social, and financial).

On 18 July 2020, the G20 Finance Ministers and Central Bank Governors Meeting highlighted the key role of the multilateral platform SOURCE in enabling "a systemic transition to the digitalization of infrastructure project development and data collection as part of advancing the work related to the QII principles."[15] This reflects the high achievement of the multilateral development banks in engaging in common activity with shared resources and joint leadership for members around the world.

In addition, on 21 May 2023, the G7 Partnership for Global Infrastructure and Investment (PGII) recognized SOURCE[16] as an important tool to improve infrastructure projects' quality, standards, and governance. "G7 also confirm the important role of other multilateral tools which improve quality, standards and governance of infrastructure projects such as SOURCE, G20 Compendium of Quality Infrastructure Investment Indicators, and the Debt Management and Financial Analysis System (DMFAS)," it is read on the communiqué.

---

[15] G20 Infrastructure Working Group. 2020. G20 Riyadh InfraTech Agenda: Background.
[16] Factsheet on the G7 Partnership for Global Infrastructure and Investment. https://www.g7hiroshima.go.jp/documents/pdf/session1_01_en01.pdf.

Integration Case Studies

Integration of SOURCE in the Philippines

Context

In 2019, the Public–Private Partnership Center (PPP Center) of the Philippines requested support from the Asian Development Bank to adopt SOURCE to serve as the agency's Project Information Management System. The Sustainable Infrastructure Foundation was mobilized under a technical assistance from the Asia Pacific Project Preparation Facility. As part of this assignment, key features were developed to transform SOURCE into an operational management system, including SOURCE's timeline, document management section, user home page reports, team member management by project status, organizational access, delegated administration interface, etc.

Rationale for Requesting SOURCE Integration

- Need for a project information management system
- Lack of interagency coordination
- Lack of accountability among agencies
- No practical solution to manage knowledge and lessons learned within organization

Integration Features

- **Policy workstream:** Support to ensure compliance with circular on "Cloud First Policy" with regard to public data management, agreement on a memorandum of understanding for the operation period,
- **Process workstream:** Development of country-specific timeline for public–private partnership (PPP) projects (18 main branches based on project origination, presence of government support, applicable legal framework and tender process) for a total of 784 predefined tasks. Deployment of knowledge management system to manage lessons learned.
- **Information technology (IT) workstream:** Development and implementation of backup application programming interface (API) and Country and Administration Interface. Discussions ongoing to ensure interoperability between SOURCE and the National Economic and Development Authority's IT tools via SOURCE's API.

INTEGRATION FEATURES

Country
Philippines

Focal Agency
Central PPP Unit (PPP Center)

Other Key Agencies
- NEDA (Planning department)
- Department of Finance
- Implementing Agencies

Supporting International Financial Institution
Asian Development Bank (Asia Pacific Project Preparation Facility)

Scope
As part of first phase of deployment, PPP projects only (BT, BLT, BOT, BOO, BTO, CAO, DOT, ROT, ROO), implemented by National Government Agencies and local government units.

- **Implementation:** Training of 20+ staff in 7 government agencies, registration of all identified organizations and users on platform, and preparation and update of complete and bespoke user guide (200+ pages).

Timeline

Needs Assessment	Development	Testing and Piloting	Rollout
Q1 2019	Q3 2019	Q2 2020	Q3 2021

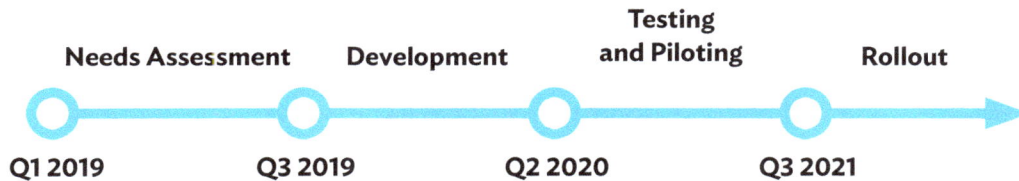

Lessons Learned

(i)   **Need for customization:** The integration of SOURCE in the Philippines was the first integration initiative, and the needs assessment phase identified all new features now being deployed in all countries. The need for more customization (country-specific settings, custom timeline, adaptation of templates, etc.) to ensure proper implementation of the tool was confirmed during the needs assessment.

(ii) **Political commitment:** The adoption of SOURCE requires an effort from all involved agencies to manage change in their respective processes. It is therefore key that the implementation of the platform is led by a well-identified agency, with the strong political support from a central government agency (e.g., presidency, ministry of finance). Such a material commitment is now compulsory for the integration process (submission of letter of interest).

(iii) **Integration timeline:** Integration was delayed multiple times due to several revisions to the platform requested by the PPP Center throughout development. This was particularly critical as the assignment included the development of major features of SOURCE. Now that the need for additional functionalities has plateaued, the risk of delay has been reduced. In addition to mitigate this risk, Sustainable Infrastructure Foundation (SIF) has developed a more thorough needs assessment process that is validated by the government prior to starting development.

Feedback

The PPP Center "adopted SOURCE to bring transparency and accountability across all government agencies involved in the Philippines' project cycle, while benefiting from a robust and fully customized project information management system and industry's best practices."

Ferdinand Pecson
Executive Director of the PPP Center of the Philippines

Reference contact:
Feroisa Francisca T. Concordia
Director
PPP Center of the Philippines

Integration of SOURCE in Uzbekistan

Context

In 2019, the Ministry of Finance of Uzbekistan requested support from the Asia Pacific Project Preparation Facility to develop an integrated IT solution to support the development and operationalization of the PPP Development Agency. The objective was to develop an integrated IT solution to support the preparation of PPP projects and their effective promotion through a centralized website for both national agencies and potential investors. The technical solution was built on the full integration of SOURCE, connected to a standalone PPP website to facilitate publication of projects and information,[17] and increase visibility of Uzbekistan's PPP framework. SOURCE's public pipeline was entirely redesigned for this purpose.

Rationale for Requesting SOURCE Integration

- Need for a solution to enforce PPP project development procedure under new legislation among implementing agencies
- Poor capacity and lack of experience in PPP project development of implementing agencies
- Need for an online platform to showcase and increase the visibility of Uzbekistan's PPP framework and opportunities to a wider marketplace

INTEGRATION FEATURES
Country Uzbekistan
Focal Agency PPP Development Agency
Other Key Agencies • Ministry of Finance • Implementing Agencies
Supporting International Financial Institution Asian Development Bank (Asia Pacific Project Preparation Facility)
Scope As part of first phase of deployment, SOURCE will be used for PPP projects only.

Integration Features

- **Policy workstream:** Support in the definition of the process to initiate projects into SOURCE and identification of the role of each agency. Discussions on the institutionalization of SOURCE to facilitate its deployment and enforce its use across all implementing agencies.
- **Process workstream:** Development of country-specific timeline for PPP projects (12 main branches based on project origination, project threshold [capital expenditure], and tender process) for a total of 272 predefined tasks. Mapping of SOURCE questions against Uzbekistan's Project Concept Form structure to enable an extract of SOURCE's data at appropriate local format via the newly developed "project assessment" module of SOURCE.
- **IT workstream:** Development of public PPP website and deployment of SOURCE's public pipeline on the website, with a filter to only display Uzbek projects.
- **Implementation:** Organization of series of trainings for 20+ staff from PPP Development Agency and implementing agencies. Organization of train-the-trainer session and specific training to launch pilot projects and upload data on SOURCE templates.

---

[17]   See http://pppd.uz/pipeline.

Timeline

Needs Assessment	Development	Testing and Piloting	Rollout
Q4 2019	Q2 2020	Q4 2020	Q3 2021

Lessons Learned

(i) **Importance of a well-defined institutional framework:** One key feedback from the initial piloting phase was the need for PPP Development Agency of a legal or regulatory framework to enforce the use of the platform as part of the project cycle across implementing agencies to ensure effective and sustainable deployment of SOURCE.

(ii) **Potential of the project assessment tool in the framework of SOURCE's integration:** In 2021, the Sustainable Infrastructure Foundation deployed a new feature on SOURCE, the "project assessment" module, allowing the gathering of a selection of data points from the templates into customizable categories and using it to derive an evaluation or a score (optional). While this module was initially designed to integrate internationally recognized standards or multilateral development banks' evaluation tools (the module was first used for the Asian Development Bank's GRIS evaluation method), it proved extremely relevant to facilitate the generation of bespoke reports, or implement national prioritization/ screening methodologies as part of the integration of SOURCE. This provides a strong incentive for users to upload data onto the platform.

(iii) **Importance of a centralized public window to increase visibility.** The identified solution combing SOURCE with a public website, including macro information related to the PPP framework, provides a very complementary solution, for both the government and investors who have access to standardized, reliable, and up-to-date information in one single location.

Feedback

The PPP Development Agency is leading the integration of SOURCE "to provide better standardization and predictability of public-private partnership processes as part of the enforcement of Uzbekistan's new PPP law."

Golib Kholjigitov
Former Deputy Minister of Finance and
Head of the Public–Private Partnership
Development Agency of Uzbekistan

Integration of SOURCE in Ecuador

Context

In 2019, the Ministry of Environment and Water and the Ministry of Economy and Finance of Ecuador requested support from the EUROCLIMA+ program to adapt SOURCE to Ecuador's needs as part of the Advisory Mechanism on National Determined Contribution Financing in Latin America. The technical assistance aimed to achieve the following:

(i) Ensure SOURCE template compatibility with EUROCLIMA+ requirements for nationally determined contributions.
(ii) Integrate the use of SOURCE in Ecuador by adjusting the project preparation timeline according to national processes and milestones and to connect SOURCE with the existing IT tools.
(iii) Facilitate the use of SOURCE by agencies involved by providing training and technical assistance.

Rationale for Requesting SOURCE Integration

- Need for a project information management system including nationally determined contributions
- Lack of national infrastructure project pipeline including nationally determined contribution monitoring
- Need for a PPP projects registry and to enforce new regulations and processes
- Lack of interagency coordination and project pipeline at the subnational level
- No practical solution to manage knowledge and lessons learned within the organization

Integration Features

- **Policy workstream:** Systematize the use of SOURCE for PPPs and public investment infrastructure projects involving nationally determined contributions at the national and subnational levels by including the use of the platform in the internal processes.
- **Process workstream:** Development of country-specific timeline for PPP and concession projects as well as for public investment for national and subnational government agencies based on project origination, applicable legal framework, and tender process). Integration in SOURCE project assessment functionality of three national project evaluation methodologies.
- **IT workstream:** Development of a dedicated climate section for all templates with the support of FELICITY. Mapping of three existing systems in process (public investment, planification, and national development bank) to facilitate SOURCE interoperability.
- **Implementation:** To date, training of 50+ staff in 4 government agencies, registration of all identified organizations and users on platform, and preparation and update of complete and bespoke user guide in Spanish.

INTEGRATION FEATURES

Country
Ecuador

Focal Agency
Ministry of Economy and Finance

Other key Agencies
- Ministry of Environment
- National Development Bank
- Implementing Agencies

Supported by:
National integration:
   EUROCLIMA+ financed by the European Union
Subnational trainings of three cities:
   FELICITY (GIZ and European Investment Bank)

Scope
PPP and Public Investment infrastructure projects at national and subnational levels.

Timeline

Needs Assessment   **Development**   **Testing and Piloting**   **Rollout**

Q1 2020   Q2 2021   Q3 2021   Q4 2021

Lessons Learned

(i) **Importance of multilateral development bank coordination:** The integration of SOURCE took place around national elections, and the coordination of the integration activities with the Inter-American Development Bank's country office was a key factor in the successful implementation of SOURCE throughout this period, and ensuring continuity of political support for the initiative.

(ii) **National and subnational approach:** The integration of SOURCE and definition of timeline were conducted in parallel at the national and subnational levels. Two piloting initiatives were launched together and fed on each other's results. The outcome of the integration process is therefore a framework adapted to the two levels of governance, for both public–private partnerships and traditional procurement. Good coordination with the central government for the deployment of SOURCE sub-nationally is key to its efficient and sustainable integration.

(iii) **Online integration.** The assignment started with the pandemic of COVID-19 in the second quarter of 2020, and the Sustainable Infrastructure Foundation's (SIF) team could not travel for the entire period of the assignment. All workshops and discussions were therefore conducted at distance, via videoconferences and exchanges of emails. SIF developed new procedures to facilitate this new approach to take into consideration the fatigue that multiple videoconferences generate. The role of the local coordinator, based in Quito, was extremely important to maintain the momentum and conduct initial trainings in Spanish.

Feedback

"SOURCE strengthens the capacity of implementing agencies by facilitating access to information on best international practices, guarantee the alignment of projects with climate, environmental, social and governance aspects, improve the implementation and monitoring of PPP processes and promote projects to attract the private sector and investors."

Vanessa Medina
Director
PPP Technical Secretariat

Jeanine Corvetto
Infrastructure Advisor in Ecuador
US Department of the Treasury

Reference contact:
Monica Chavez
Advisor
Ministry of Finance, Ecuador

Integration of SOURCE in Ukraine

Context

In 2019, Ukraine was identified as the first pilot country among European Bank for Reconstruction and Development (EBRD) countries of operation, where the integration of SOURCE will take place at the national level. Ukraine has been going through systemic reforms to build a coherent and effective institutional and legal environment for implementation of public–private partnerships (PPPs) and overall improvement of public infrastructure delivery. The integration of SOURCE has been conducted in two phases: The first phase conducted the integration of SOURCE for PPP and concession projects only, and the second phase included connection of SOURCE with Ukraine's e-procurement platform and the rollout of the tool to all infrastructure projects. Due to the Russian invasion of Ukraine, the SOURCE integration process has been paused. However, the SOURCE platform will play a key role in Ukraine's reconstruction plans. SOURCE will enable Ukrainian authorities to structure projects and road show its needs to multilateral development banks, donors, and investors.

Rationale for requesting SOURCE integration

- Need for an end-to-end IT platform (in complement to the e-procurement system in operation prior to the integration of SOURCE) to support the institutional reform conducted with the support from the EBRD
- Digitalization of the full project cycle to reinforce standardization and transparency of processes
- Capacity strengthening of implementing agencies, particularly at subnational level

Integration Features

- **Policy workstream:** (i) Review of primary and secondary legislation in Ukraine potentially conflicting with SOURCE implementation and recommendations on regulatory amendments and (ii) review of compliance of SOURCE templates and milestones with requirements of the Ukrainian legislation.
- **Process workstream:** Identification and mapping of current processes within the project and development of a road map for optimization, based on the use of the SOURCE platform. SIF also assessed gender and inclusion entry points as well as inclusive procurement processes.
- **IT workstream:** Restructuring of the entire SOURCE data structure to enable the alignment of SOURCE API with Open Contracting Data Standard to facilitate interconnectivity with Ukraine's e-procurement platform.

INTEGRATION FEATURES

Country
Ukraine

Focal Agency
PPP Unit of Ministry of Economic Development and trade

Other Key Agencies
- Ministry of Infrastructure
- PPP Management Unit (SPILNO)
- Other Procuring Authorities

Supporting International Financial Institution
European Bank for Reconstruction and Development

Scope
As part of the first phase of implementation: PPP and concession projects. In second phase, all infrastructure projects.

Timeline

Needs Assessment	Development	Testing and Piloting	Rollout
Q1 2020	Q3 2020	Q3 2021	Q4 2021

Lessons Learned

(i) **Importance of data standardization:** The integration of SOURCE in Ukraine has required the restructuring of SOURCE's templates for the platform to collect data on contractual management and procurement process. This was done to standardize data collection and align SOURCE with the Open Contracting Data Standard to facilitate the use of SOURCE's API by all platforms using the same standard (e-procurement platforms, CoST transparency dashboards, etc.). SOURCE has a wider role to play to extend the scope of the Open Contracting Data Standard to the whole project cycle, from definition to implementation.

(ii) **API strategy:** The integration of SOURCE in Ukraine confirmed the relevance of the API strategy that had been decided with the SOURCE Council. Rather than trying to capture all the digital needs from governments throughout the project cycle into SOURCE, the most efficient approach proved to maintain the scope of SOURCE's services unchanged and develop a comprehensive IT system by connecting SOURCE to third-party applications (either preexisting or developed as part of the integration) via SOURCE's API.

(iii) **Importance of policy approach:** As part of the integration in Ukraine, the EBRD supported a wider legal reform that allowed including the use of SOURCE in the regulation. This proved a key factor in successful integration and was replicated in other countries.

Feedback

"SOURCE is one of the best project management tools I have used to prepare well-structured projects in PPP, concessions and standard procurement."

Taras Boichuk
Head of SPILNO PPP Office
Ukraine

Reference contact:
Eliza Niewiadomska
Senior Counsel
EBRD, Public Procurement, Digital Economy, Open Government

APPENDIX
SOURCE's Major References

Joint communication to the European Parliament, the Council, the European Economic and Social Committee, the Committee of the Regions and the European Investment Bank—Connecting Europe and Asia—Building Blocks for an EU Strategy, 2018.

> "The support of multilateral development banks is essential in implementing the G20 'Roadmap to Infrastructure as an Asset Class', and a wider adoption of infrastructure management platforms would help improve the implementation of projects. For example, for the management platform SOURCE."

> See the website of the Sustainable Infrastructure Foundation: https://public.sif-source.org/. https://eeas.europa.eu/sites/default/files/joint_communication_-_connecting_europe_and_asia_-_building_blocks_for_an_eu_strategy_2018-09-19.pdf.

Organisation for Economic Co-operation and Development (OECD) Reference Note on Environmental and Social Considerations in Quality Infrastructure, June 2019.

> "Through the implementation of digital smart contracts and digital due diligence documentation, greater social and environmental outcomes could be validated and advanced. A digital approach would expand transparency and speed of information for all of the stakeholders associated with quality infrastructure projects. Additionally, the use of digital tools and knowledge platforms, such as SOURCE, should be scaled-up in countries, especially in those with lack of expertise in developing quality infrastructure projects." (p. 10)

> https://www.oecd.org/g20/summits/osaka/OECD-Reference-Note-on-Environmental-and-Social-Considerations.pdf.

OECD/IMF Reference Note on the Governance of Quality Infrastructure Investment, June 2019.

> "Infrastructure policy should be based on data to inform decision making and stakeholder engagement, as identified in Principle 6.4 of the G20 Principles for QII, which includes the use of digital technology. Governments should put in place systems that ensure a systematic collection of relevant data and institutional responsibility for analysis, dissemination, and learning from this data. Relevant data should be disclosed to the public in an accessible format and in a timely fashion. Effective monitoring and evaluation frameworks should be designed and integrated into the decision-making process. To this effect, digital and open-source information will help enhance infrastructure investment policies, improve the management of public resources, promote transparent and accountable projects, and improve ex-post evaluation. For instance, the multilateral platform SOURCE provides project developers with a means to manage project information and preparation throughout the infrastructure project cycle and helps disseminating related materials." (p. 9)

> https://www.mof.go.jp/english/policy/international_policy/convention/g20/annex6_5.pdf

EBRD and World Bank, MDB Infrastructure Cooperation Platform: Project Preparation Workstream—Phase II Reference Note on Project Preparation Across the Full Project Cycle, June 2019.

"This reference note recognises that there are many tools and approaches available. These provide valuable contributions that complement multilateral development banks' offering across the product cycle and we look to enhance and deepen the process of collaboration in the inclusive spirit of the systemic approach advocated by MDB shareholders. An example of this is the SOURCE platform. SOURCE has been designed to provide a standardised yet adaptive framework allowing for the integration and harmonisation of several international tools and standards, for dissemination and data collection purposes. SOURCE already integrates many internationally recognised tools and approaches and we look to enhance and deepen this process. As they further mainstream these tools into normal operations, the multilateral development banks expect a more systemic approach to emerge, while recognising, of course, the inherent uniqueness of each project in its particular market context." (p. 2)

"Lastly, the SOURCE platform, led and funded by the multilateral development banks since 2014, does not represent a reference product or tool per se, but should be rather considered as an enabling platform to deliver scale and consistency to the entire workstream presented in the document. Through its structured approach to project data and the growing adoption of the platform by national government agencies and MDBs, SOURCE would also enable the collection of standardised project data globally and the usage of MDB tools." (p. 3)

"SOURCE is the multilateral infrastructure project platform implemented by the Sustainable Infrastructure Foundation (SIF). Several MDBs (multilateral development bank), including (Asian Development Bank, Inter-American Development, European Investment Bank, World Bank, and European Bank for Reconstruction and Development), provide key inputs into SOURCE, and since 2018, the strategic and financial management of SOURCE is under the supervision of the SOURCE Council, which is composed of representatives from MDBs. SOURCE provides a structured approach to the investment cycle through sectorial templates, hereby enabling: (i) the provision of a standardised and comprehensive map of all aspects to take into account the develop of high quality, sustainable infrastructure; (ii) deliver MDB tools, reference notes and best practices to project managers at the right juncture in the decision process; (iii) monitor whether projects meet their intended outcomes and benefits during the implementation period, and (iv) collect structured and standardised project data at global scale to assess performance of projects against standards, generate analytics, and benchmarks (for example, unit costs). SOURCE has been designed as a public good, to be used by government agencies and MDBs." (p. 10)

https://www.mof.go.jp/english/international_policy/convention/g20/annex.htm

United Nations Economic Commission for Europe, Committee on Innovation, Competitiveness and Public–Private Partnerships, Working Party on Public–Private Partnerships, 2020.

"This Tool was initially developed as an Excel platform for testing purposes and is being integrated into the SOURCE platform operated by the Sustainable Infrastructure Foundation (SIF); At the request of the member States (paragraph 24, ECE/CECI/WP/PPP/2019/2), the UNECE secretariat is collaborating with the Sustainable Infrastructure Foundation (SIF) and its SOURCE software in the implementation of this methodology. SIF is not-for-profit entity funded by the Multilateral Development Banks and provides practical guidance to countries in project development." (p. 22)

https://unece.org/sites/default/files/2020-12/ECE_CECI_WP_PPP_2020_03_Rev1-en.pdf

G20 Finance Ministers and Central Bank Governors G20 Riyadh InfraTech Agenda, 2020.

> "We endorse the G20 Riyadh InfraTech Agenda, which promotes the use of technology in infrastructure, with the aim of improving investment decisions over the lifecycle, enhancing value for money of infrastructure projects, and promoting quality infrastructure investments for the delivery of better social, economic and environmental outcomes." "The multilateral platform SOURCE, to enable a systemic transition to the digitalization of infrastructure project preparation and data collection as part of advancing the work related to the QII principles." (p. 7)

https://cdn.gihub.org/umbraco/media/3008/g20-riyadh-infratech-agenda.pdf

French Presidency, Declaration of the Summit on the Financing of African Economies, Paris, 2021:

> "Supporting capacity development for planning and preparation of key infrastructure projects and foster the emergence of bankable projects, through promoting the deployment of the multilateral platform SOURCE for sustainable infrastructure project preparation, jointly led and funded by multilateral development banks." (p. 8)

https://www.elysee.fr/index.php/en/emmanuel-macron/2021/05/18/summit-on-the-financing-of-african-economies-1

Asian Development Bank, Supporting Quality Infrastructure in Developing Asia, July 2021.

> "SOURCE is a customizable, secure web application for DMC (developing member country) project management, planning, processing, assessment, and overall preparation. Its project management functions include project pipelines, documents, portfolio management, and monitoring dashboards for project preparation, all of which can be tailored to a DMC's regulatory regime and project approval processes. SOURCE project preparation templates provide guidance on managing projects even where the private sector is involved. The SOURCE templates are in accordance with global standards like the (International Finance Corporation) performance standards on environment, social, and governance, the Global Infrastructure Facility Project Preparation Readiness Assessment, and the Association of Project Managers Group PPP Certification Guide's gateway process. The platform also integrates internationally recognized knowledge products such as the (International Monetary Fund) and the World Bank PPP Fiscal Risk Assessment Model that analyzes fiscal risks of infrastructure projects, the Global Infrastructure Hub PPP risk allocation tool, and the United Nations Environment Programme (UNEP) Principles for Positive Impact Finance.

> SOURCE was cited in the G20's InfraTech agenda as a means for countries to enable a systemic transition to the digitalization of infrastructure project preparation and data collection in relation to QII principles. SOURCE helps governments improve investment decisions over a project's life-cycle, enhance value for money, and promote quality infrastructure investments to achieve better economic, social, and environmental outcomes. As a multilateral and online platform, SOURCE holds great potential to digitize and disseminate ADB tools and diagnostic instruments related to infrastructure governance and sustainability, such as the Climate Risk Management Framework, Procurement Risk Assessment, and Value for Money Guidance for Procurement, among others (Table 2). This will enable ADB to expand its knowledge sharing activities and increase DMC access by having tools and best practices available in one place along with knowledge products from other multilaterals." (p. 24)

https://www.adb.org/sites/default/files/publication/715581/supporting-quality-infrastructure-asia.pdf

References

ADB. 2009. Establishment of e-Systems in Support of Infrastructure Finance in Asia. Available: https://www.adb.org/projects/documents/establishment-e-systems-support-infrastructure-finance-asia.

EBRD and World Bank Group. 2019. MDB Infrastructure Cooperation Platform: Project Preparation Workstream—Phase II Reference Note on Project Preparation across the Full Project Cycle. Washington, DC. Available: https://www.mof.go.jp/english/international_policy/convention/g20/annex6_2.pdf.

G20 Infrastructure Working Group. 2020. [Online]. G20 Riyadh InfraTech Agenda: Background. Available: https://cdn.gihub.org/umbraco/media/3008/g20-riyadh-infratech-agenda.pdf.

G20 Principles for Quality Infrastructure Investment. 2019. [Online]. Available: https://www.mof.go.jp/english/international_policy/convention/g20/annex6_1.pdf.

G20 website. 2019. [Online]. Available: https://g20.org/en/about/Pages/whatis.aspx.

HSBC. 2021. [Online]. Available: https://www.sustainablefinance.hsbc.com/sustainable-infrastructure/fast-infra-a-public-private-initiative.

HSBC. 2021. Fast-Infra: A Public–Private Initiative. [Online]. Available: https://www.sustainablefinance.hsbc.com/sustainable-infrastructure/fast-infra-a-public-private-initiative.

Investor Leadership Network website. 2022. [Online]. Available: https://www.investorleadershipnetwork.org/en/.

PPP Ukraine website. 2022. [Online]. Available: http://pppd.uz/pipeline.

Sustainable Infrastructure Foundation. 2022. Available: https://public.sif-source.org/.

United Nations. 2015. Transforming Our World: The 2030 Agenda for Sustainable Development. Available: https://sustainabledevelopment.un.org/post2015/transformingourworld/publication.

United Nations International Computing Centre website. 2022. [Online]. Available: https://www.unicc.org/.